Isambard Kingdom Brunel
Engineer Extraordinary

Few men have contributed more to civil engineering than Isambard Kingdom Brunel. His wide-ranging achievements included the Thames Tunnel, the Great Western Railway, and the *Great Eastern* steamship, which laid the first trans-Atlantic telegraph cable in 1866.

He was indeed an extraordinary man, and his ideas matched his spirit. His design for the Clifton Suspension Bridge was condemned by Thomas Telford as impossible to build. His tunnel under Box Hill was said to be doomed to collapse. Both of them stand to this day. But his fierce determination and an incredible capacity for work led him to an early grave.

There were disasters and failures, too. He was nearly killed in 1827, when the Thames Tunnel flooded, and again in 1838, when the *Great Western* steamship caught fire. He lost his battle with George and Robert Stephenson over the broad-gauge railway, and his attempt to build an atmospheric railway was a complete failure. Yet his successes were outstanding.

Brunel's career is central to the study of transport and communications in the nineteenth century, and in this survey David and Hugh Jenkins record each stage of his remarkable life. The book includes a date chart, glossary, reading list, and index, and there are more than fifty illustrations.

For Rhoda, and in fond memory
of "Jack who was a railwayman too"

The *Great Eastern* leaves Southampton Water bound for New York in June 1860

Pioneers of Science and Discovery

Isambard Kingdom Brunel
Engineer Extraordinary

David and Hugh Jenkins

WAYLAND PUBLISHERS LTD

Other books in this series

*Jacket photographs courtesy of
the Science Museum, London, and
British Railways.*

SBN 85078 210 4
Copyright © 1977 Wayland Publishers Limited
First published in 1977 by
Wayland Publishers Limited
49 Lansdowne Place, Hove, East Sussex BN3 1HF
Second impression 1980
Printed and bound in Great Britain
at The Pitman Press, Bath

Contents

DAVID AND HUGH JENKINS are identical twin brothers. They were born in Cardiff in 1934, and spent their formative years train-spotting on the Western Region in Wales. They both took a degree in English at University College, Cardiff, where David Jenkins went on to do a research degree in Medieval Religious Drama. He is now a Lecturer in Education at the Open University, and is on loan to the University of East Anglia, working in computer-assisted learning.

Hugh Jenkins pursued a career in railway management after leaving university, and he is now Director of Studies at the British Transport Staff College in Woking.

List of Illustrations

Above Marc Brunel's designs for a
new Wall Street bank, drawn up
while he was Chief Engineer of
New York

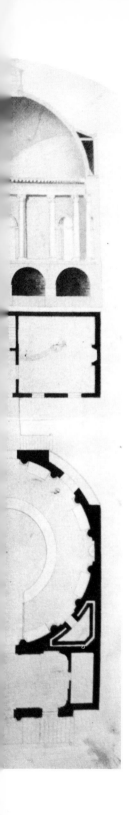

1 Marc Brunel

The name Brunel is one of the most famous in the history of civil engineering. To most of us it means Isambard Kingdom Brunel, builder of great bridges and steamships, and pioneer of the Great Western Railway. There was, however, another famous Brunel: Isambard's father, Marc.

Marc Isambard Brunel was born in Hacqueville, in Northern France on 25th April, 1769. His father, a prosperous farmer, was determined that he should become a Catholic priest. Marc, however, was not interested in the priesthood. He did not enjoy studying Latin, and preferred to spend his time hanging around the local carpenter's shop, or sketching buildings and machinery. When he eventually persuaded his father that he was not suited to a religious training he was allowed to travel to Rouen, where he could study maths and drawing, and train for the French navy. In Rouen he met and fell in love with an English girl called Sophia Kingdom.

The year was 1793 and the French Revolution was at its height. Marc was a well-known and rather noisy Royalist, so when the Republicans seemed about to grasp power he became a wanted man. Following a sudden crisis, he fled to New York aboard the good ship *Liberty*, with a forged passport.

The enterprising Marc soon made a name for himself in New York. He took up work as a land surveyor and soon worked his way up rapidly and became Chief Engineer of New York. He took United States

9

citizenship, and a rosy future seemed to stretch before him. But he was a restless man and keen to move on. He knew that his beloved Sophia had journeyed to England, and he wanted to join her there. He soon had the chance to do so. One evening at a dinner party, he heard of the firm of Taylors of Southampton, who at that time held a monopoly for the manufacture of ships' blocks. (These are the wooden pulleys used to raise the sails of sailing ships). Blocks were not only expensive to make, they also had to be made very accurately. Marc saw where he could improve on Taylors' design. He took his chance and came to England in 1799, and was re-united with Sophia.

Left Block used on Telford's suspension bridge over the Menai straits

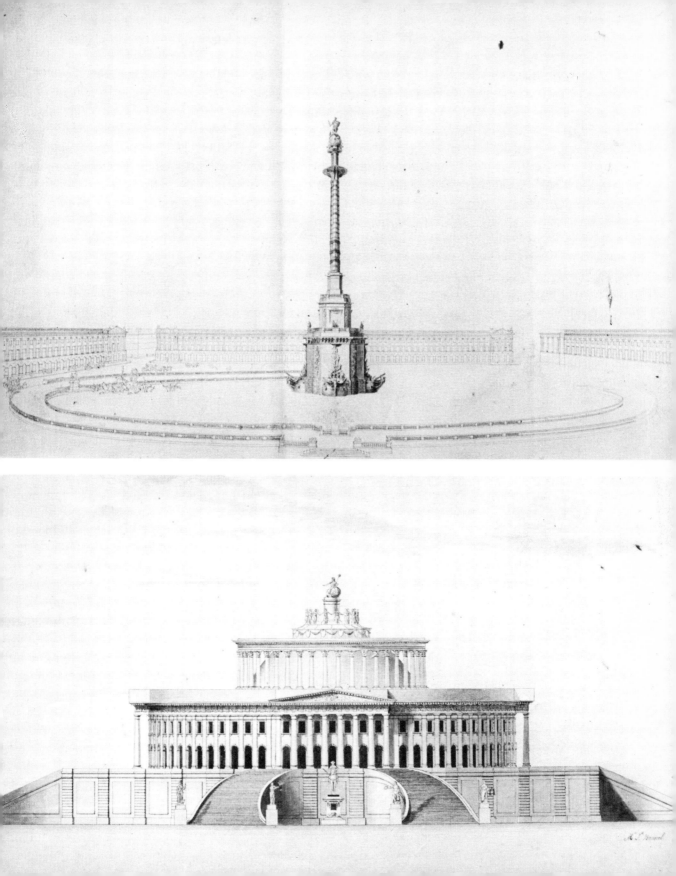

Soon after he arrived in England, Marc Brunel met Henry Maudslay, who was shortly to found the most famous engineering firm of the nineteenth century, Maudslay, Sons and Field. At that time Maudslay was working with a single assistant in a little workshop in Wells Street, off Oxford Street, in London. He was immediately interested in Brunel's ideas and drawings for machines to make ships' blocks. The only problem now was to find enough money to set up the machines. Sophia's father approached Taylors of Southampton on Marc's behalf. He suggested that Marc's ideas might be introduced into Taylors' business. But Taylors were unwilling to share the market, and refused to co-operate. Then Sir Samuel Bentham, Inspector-General of Naval works, took a personal interest in the project. He liked the plans put forward by Brunel and Maudslay, and told them to go ahead.

Over the next six years the block-making machines were made up. Using Marc Brunel's designs, Henry Maudslay built what were the first machines to replace human workers. They could produce blocks more accurately and more quickly than the teams of men who had always been needed to make such equipment. The partnership was a great success.

On 1st November, 1799, Marc Brunel and Sophia Kingdom had been married, and by 1806, the year the block-making machines began operation, they had two daughters, Sophia and Emma. On 9th April of this year they had a son. He inherited his first name from his father. His middle name came from his mother's maiden name. He was christened Isambard Kingdom Brunel.

Isambard was a lively and energetic child, and brought to both work and play the enthusiasm and determination which later played such an important part in his career as an engineer. As we shall see, he surpassed even his father in his constant search for new ideas and projects.

Above Sir Samuel Bentham (1757–1831), Inspector-General of Naval Works and the man who recognized the value of Marc Brunel's block-making designs

Below The famous "boot" caricature of the Duke of Wellington, 1827

After his success with the block-making machines, Marc Brunel was now looking for new schemes to work on. He built a factory to mass-produce army boots, and set up a large steam-driven sawmill at Battersea. This time, however, luck was against him. On the night of 30th August, 1814, the sawmill burned down in a fire that lit the London sky. Then, in the sudden peace following the Battle of Waterloo, the army no longer needed boots, and Marc was left with a useless factory. After these two disasters he went out of business and soon fell into debt. On the 14th May, 1821, Marc Brunel was arrested for debt and locked up in the grim King's Bench prison. The courageous Sophia was allowed to go with him. A prison visitor gave us a glimpse of their life there, recalling "the small room, in one corner of which was Brunel at a table littered with papers covered with mathematical calculations, while, seated on a trestle bed in the opposite corner, sat his wife mending his stockings". Several months passed before the Duke of Wellington was able to raise £5,000 and secure his release.

Despite his financial difficulties, Marc Brunel still managed to give his son a good education. The young Isambard was revealing a talent for drawing and had grasped the principles of geometry at the astonishingly early age of six. He was sent to Dr. Morell's boarding school at Hove, where he showed the same love of sketching that his father had shown in Rouen. He also proved himself to have acute powers of observation. While at school, he amazed his friends by telling them that a new building being built opposite the school was in danger of sudden collapse. He was right in his prediction. He was already proving his father's belief that successful engineering required not only complex mathematical calculations, but also an engineer's intuition. Thus, both Brunels were able to come to an almost instant judgement about the stability of a physical structure.

The next part of Isambard Brunel's education was in France, first in his father's native Normandy, then in the Lycée Henri-Quatre in Paris. Marc Brunel probably chose this school with its famous mathematics teachers in mind. After this, Isambard was sent to Paris to serve an apprenticeship under the famous Abraham Louis Breguet, the best maker of chronometers, watches and scientific instruments of his time. Breguet was a very good craftsman, painstaking and accurate. The impression he made on Brunel lasted for the rest of his life.

When he returned to England at the age of sixteen Isambard Brunel was introduced to Henry Maudslay, his father's old colleague, and spent many absorbing hours in his workshops. Maudslay was as good an engineer as Breguet was a watchmaker. From him Isambard Brunel learned a great deal.

Below The Lycée Henri-Quatre in Paris in the late eighteenth century

Above Isambard Kingdom Brunel
at the time of the Great Western
Railway Survey – from a portrait
by John Horsley, RA

By the age of eighteen, Isambard Brunel had finished his education and apprenticeship. His father decided that he was ready to tackle his first engineering project. Marc had in his mind an ambitious plan, one that was to give Isambard a difficult start to his career – a tunnel under the River Thames.

Although Isambard Brunel had the advantages of his famous father and an unusual education, he was in many ways similar to his self-made father. Both were self-willed, tough and determined. Both put mistakes behind them easily and could keep their mind on the job in hand. Perhaps Isambard's self-confidence was more openly displayed. At times it came close to arrogance or simple conceit. His private journal confesses his "self-conceit and love of glory". He occasionally indulged in pointless displays. Once, when passing a stranger at night, he sat up as tall as possible on his little pony to look physically impressive. Later he recognized his ability to awe people, and would allow himself bursts of scorn and anger. His father by contrast was a calmer man and some people found the grand old man more impressive than the young Isambard, with his brooding genius and blazing black eyes. Charles MacFarlane, who knew both men, wrote of them, "I had liked the son, but at our first meeting I could not help feeling that his father far excelled him in originality, unworldliness, genius and taste." Both men attracted followers, even disciples. Both, too, were able to set up rewarding partnerships with equals, but hesitated to share direct responsibility.

They were desperately hard-working and put all their energy into current projects. Both were able to shrug off tiredness and disappointment. Isambard Brunel's relentless journeying on cold, overnight stage coaches astonished those who could only work effectively after a good night's sleep. But while Marc was able to grow old gracefully, Isambard worked himself into an early grave.

THAMES IRON WORKS & SHIP-BUILDING Cº BLACKWALL

2 Tunnel under the Thames

When Marc Brunel formed the Thames Tunnel Company in 1824, it was by no means the first time that engineers had attempted to drive a tunnel under the river. In 1802, Robert "the Mole" Vazie, a Cornish mining engineer, had put a daring scheme forward. Three years later the Thames Archway Company was formed. It was given power, by an Act of Parliament, to build a tunnel according to Vazie's plan. This plan was in two parts. First he had to dig a driftway beneath the river, only three feet wide at its base and five feet high (0.9 x 1.5 m), narrowing towards the top. The second, and more difficult part of Vazie's plan, was to construct a full-sized tunnel over the top of the driftway. This would allow the driftway to act as a drain for the main tunnel. Many people were unimpressed by Vazie's plan, for he was disturbingly vague about how he would tackle the biggest problem of all. This was how to advance the "face" of the main tunnel through shale and sand without flooding the workings and drowning the miners. This problem also confronted Marc and Isambard Brunel.

Behind the advancing face of a tunnel, sections could quickly be reinforced. Narrow driftways often used pit props for this purpose. A larger tunnel, however, unless driven through solid rock, needed something like a bricked arch to support the roof and walls. Even so, this always left an unprotected area between the face itself and the reinforced tunnel. If the river came through into the workings, it did so at

Left The Thames Tunnel shaft under construction

17

this point. This was the danger of tunnelling through a mixture of rocks beneath water. But Vazie did not get far enough to experience these problems. His driftway ran into a bed of shale, always loose and treacherous, particularly under water. The river seeped through the shale overwhelming the pumps and flooding the workings. Then Richard Trevithick, another Cornish engineer, was sent for. He was offered one thousand pounds to finish the driftway. "I think this will be making one thousand pounds very easily," he declared, and estimated that the task would take him six months. It was an optimistic statement made by an optimistic man. His claim was wide of the mark. The high tide of 26th January, 1808 put pressure on the ceiling of the tunnel over the face, which was edging forward hopefully through saturated sand and shells. Suddenly the river burst in, and Trevithick found himself up to his neck in water. Luckily he was able to escape from the flooded tunnel. He hired a Thames barge and dropped bags of clay overboard in an attempt to plug the hole. If this had proved successful, the pumps would have quickly cleared the workings of water. But the river again broke through, and once more water poured into the driftway through breaches in the roof. If a small driftway could not be made watertight, a full-size tunnel seemed to be impossible. It was no surprise when, in 1808, the Thames Archway Company gave up the attempt.

It was sixteen years later when Marc Brunel made his plans for a tunnel under the Thames. Like all his plans they were very grand. There were to be two parallel tunnels, each six metres in diameter, and wide enough to carry squires' coaches or city merchants' drays. The scheme was also technically clever. In order to overcome the problem of collapse at the tunnel face, he devised a huge mechanical shield to protect the workings. This novel idea was suggested to Marc Brunel by the head of the *teredo*

Above Richard Trevithick (1771–1833), the famous Cornish Engineer who was brought in to finish the driftway for the Thames Tunnel – not as easy a prospect as he first thought!

Right Cross-section of the Brunel
Tunnelling Shield

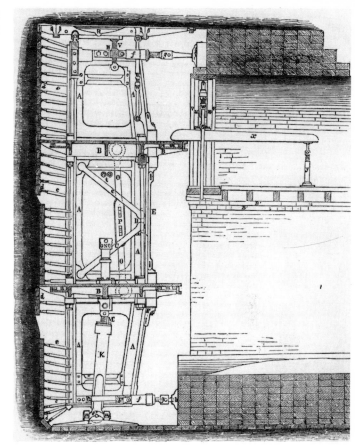

Below An oblique view of the
Tunnelling Shield in position

insect which he had seen burrowing through ships'
wooden hulls in Chatham Docks. He also moved the
site of the tunnel three-quarters of a mile upstream
from Vazie's ill-fated driftway. The tunnel was to run
"from some place in the parish of St John of Wapping
in the County of Middlesex to the opposite shore of
the said river in the parish of St Mary, Rotherhithe in
the County of Surrey, with sufficient approaches
thereto".

At the start of the project, Marc Brunel ordered a
survey of the underground rocks. The drill holes
bored as a part of this survey revealed nine separate
layers of rock beneath the bed of the river. He had
hoped to find a layer of clay thick enough to tunnel

19

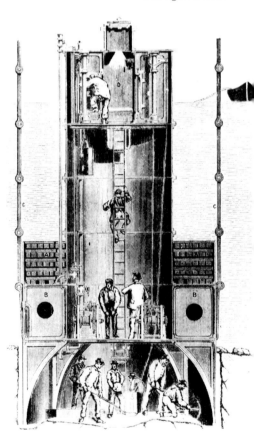

through, but the clay turned out to be only a metre thick. He then decided to tunnel beneath the clay, using it as a "roof" over the workings, even though this small amount of protection did not extend along the whole length of the tunnel. Already there were many difficulties, but Marc Brunel's "great shield" attracted public confidence and enough financial backing to go ahead. The Thames Tunnel Company was formed, with Marc Brunel as Chief Engineer. At the opening ceremony Marc laid the first brick, Isambard the second.

The first major task was to sink a shaft down to the level at which the miners could begin digging the tunnel itself. This was done in a very clever way. Marc Brunel first had a round, brick tower built. Then he ordered the ground beneath the rim of the tower to be dug out, allowing the tower to sink slowly into the ground. This "sinking tower" became one of the wonders of London. Many people came to watch the work in progress. But it wasn't only the gaping populace who paused to marvel at the disappearing tower and great iron shield. The miners themselves had never worked this way before and regarded the contraption with a mixture of apprehension and awe. Long after it ceased to be a novelty it remained a mystery.

Above The Thames Tunnel in 1827

Soon the huge, black, iron insect was burrowing its way towards Wapping. Within the shield, the face was divided into a number of compartments, arranged side by side and one above the other. It was as if each miner worked his own section as a private driftway. After only nineteen centimetres excavation, the shield would be jacked forward into its new position. Helping the miners was a gang of workers whose job was to carry the rock which had been dug out (the spoil) away from the face in skips (large baskets). This waste was hauled up the shaft by steam-engine. Even those sight-seers who at a later date would be unable to afford the entrance money to the tunnel proper, were able to watch the growing spoil heaps on the surface. In the early days of the project, Isambard Brunel, when only nineteen, had worked as a first assistant under William Armstrong the Chief Engineer. Armstrong was working under great pressure in very bad conditions and when he was taken seriously ill, Isambard took over his job.

21

As work on the tunnel advanced, Isambard became more and more involved in the enterprise until he eventually carried the main day-to-day responsibility for the work as "resident engineer". He did this for the princely sum of £200 per year. The ruthlessly hard working Isambard was often quite literally "resident", working at all hours and even sleeping on the site.

Life below the Thames was unpleasant as well as dangerous. The river itself was little better than an open sewer. Trash from the river bed would frequently seep through into the workings. Excrement was mixed freely with butchers' bones, old fragments of china and silt. At times the stench in the tunnel was horrible. Dangers to health walked hand in hand with the physical dangers. A number of miners suffered nausea, sickness and even temporary blindness.

Although the shield worked reasonably well, the tunnel was frequently passing through unstable

Below The relative position of the Thames Tunnel and the river above it

Above Section of the Thames Tunnel with the double approaches as originally planned

material, and not surprisingly the river burst in several times. Brunel's response to these breaches was the same as Trevithick's before him. He took a look at the damage from above. Indeed, Brunel nearly drowned himself in a diving bell, borrowed from the West India Dock Company, to inspect the results of one such calamity. He was a very agile young man who had more than his fair share of luck. Another lucky escape was made by Tillet, the engine man. Following one invasion of the tunnel by the murky waters of the Thames, Tillet found himself swimming about at the bottom of the shaft "like a rat . . . quite spent". Brunel descended on a rope and winched the old man up the shaft to safety.

Brunel appeared at times to enjoy the drama and the heroics of these incidents. It seemed that he was almost fearless. Taking a brave inspection party into the already-flooded tunnel, Brunel was heard to warn, "Now gentlemen, if by accident there should be a rush of water, I shall turn the punt over and prevent you being jammed under the roof, and we shall then be carried out and up the shaft." This warning wasn't heeded and non-swimmers continued to join parties punting along near the tunnel roof in three metres of water. Then a punt overturned, and one of the miners was drowned. Isambard Brunel's own feelings towards the danger were shown by an entry in his private journal following an accident on 11th January, 1828, when the face of the tunnel burst and Isambard was trapped for a time under water. Brunel appeared to enjoy it almost in the same way as you or

23

I might enjoy a horror movie. He recorded:

"I stood still nearly a minute. I was anxious for poor Ball and Collins, who I felt sure had never risen from the fall we had all had and were, as I thought, *crushed* under the great stage. I kept calling them by name to encourage them, and make them also (if still able) come through the opening. While standing there the effect was – *grand* – the roar of the rushing water in the confined passage, and by its velocity rushing past the opening was grand, *very grand*. I cannot compare it to anything, cannon can be nothing to it."

But the most fantastic spectacle had taken place one year previously, on Saturday 10th November, 1827. Brunel and his father had held a breathtakingly beautiful banquet beneath the river. The brick arches were hung with crimson, and the tables set out with gas candelabra. In the next arch a feast was put out for the workers, headed by the élite of the working gangs, the face workers themselves. But even this dazzling occasion failed to set the tone for the future.

The Thames continued to threaten progress, and on 14th January, 1828, it flooded the tunnel again, in the incident described earlier in Brunel's journal. Despite the thrill he felt at seeing the water sweeping through the tunnel, Brunel had been badly injured, and he was forced to rest.

After this incident, the financial backers of the project stopped the supply of money, and work on the tunnel came to a halt. The shield was bricked up, and Isambard felt that his father would never see the tunnel completed.

In fact, work began again in 1835, and the tunnel was opened in 1842. By this time, however, Isambard Kingdom Brunel, although suffering from his accident and a long illness, had turned his restless mind to search for greatness elsewhere.

Right The Thames Tunnel – open for business

3 *Isambard Brunel in Bristol*

Following his long illness at Bridge House, the Brunel family house in London, Isambard Brunel's parents decided that he should spend some time convalescing at Clifton in Bristol. Brunel felt at home in Clifton, the merchant's settlement on the limestone crags above the river Avon, and spent his time scrambling and sketching among the rocks by the river. From here he could see the tall ships sail up the River Avon to the inland port of Bristol. At Clifton they passed between towering limestone cliffs in a deep, wide gorge. The idea of building a bridge across this gorge had been discussed for many years and was just the kind of project that excited Brunel.

In 1753, a Bristol wine merchant had left some money in his will to start a bridge-building fund. But it was not until 1829 that a Bridge Committee was set up to do something about it. They hoped to collect money from the public, and they invited designers to draw up plans and enter them in a competition to decide the best design. Brunel set to work. He prepared a series of drawings and sent them to the Committee. But when the Committee asked the old bridge designer, Thomas Telford, to judge the designs they had a bit of a shock. He did not like any of them. The great man, by then over seventy years old, sternly warned that the gorge was too wide to span without some supports in the middle. He claimed that 600ft (183 m) was the greatest single span possible. Significantly,

Left Thomas Telford (1757–1834), the father of civil engineering, first President of the Institution of Civil Engineers, and judge of the first competition of designs for the new bridge over the Clifton Gorge

This was just the length of his own record-breaking Menai Bridge! The Bridge Committee then asked Telford to submit his own design. Although Telford had been a great designer in his day, he was too old to consider an imaginative solution to the problem of bridging the Clifton Gorge. His design featured two enormous towers rising up from the floor of the gorge. These would carry most of the weight of the bridge. Telford's oddest idea was to decorate the two enormous piers in the style known as Florid Gothic. This would disguise the supports as church towers, complete with artificial spires and pinnacles. The gorge would be split into three, like the nave and aisles of a gothic church. Brunel did not hesitate to criticize Telford's design. He wrote a letter to the Bridge Committee in which he insisted that the gorge could be bridged without supporting piers merely by spanning the whole width of the gorge. He added:

28

"As the distance between the opposite rocks was considerably less than what had always been considered as within the limits to which Suspension Bridges might be carried, the idea of going to the bottom of such a valley for the purpose of raising at a great expense two intermediate supports hardly occurred to me."

In order to overcome the problem of Telford's dislike of all designs except his own, a second design competition was arranged in 1830. Isambard Brunel entered another design for this new competition and to his delight the judges declared him the winner.

In June, 1831, work started on the Brunel bridge, soon to become "the ornament of Bristol and wonder of the age". But before this prediction could come true the work was stopped because of lack of money. When, in 1836, work resumed, a flimsy basket hanging below a bar was provided as a means of crossing the gorge. It trundled across the gorge, hauled across by ropes. One day, Brunel, the daring adventurer, took a ride in this basket. When it jammed in the middle, he impulsively climbed up from the basket to clear the roller. There were few spectators who did not miss a heartbeat. He could easily have been killed. But all through his life Brunel exposed himself to danger in this sort of way.

After Brunel's death in 1859, the Institution of Civil Engineers decided to complete the bridge in his memory. Many of his friends and opponents contributed money for this venture. In December, 1864, the Clifton Suspension Bridge was opened, five years after Brunel's death. Although there were small changes to the original plan, the principles of design were those that Brunel had put forward many years before. The bridge stands today and is still considered an impressive structure. It is a fitting memorial to the adventurous spirit of the young engineer.

Below Design for the gateways to the proposed suspension bridge over the River Avon at Clifton

The suspension bridge at Clifton,
nearing completion in 1863, after
Brunel's death

4 Great Western Railway

His early work on the Clifton Bridge had earned
Brunel great admiration from the Bristol merchants,
and they invited him to redesign the floating docks at
Bristol. While he was working on this, he heard of a
scheme to build a railway from Bristol to London.
The committee in charge of pursuing this idea
wanted to appoint an engineer who would spend as
little of their money as possible in surveying the line.

Still anxious to make his name in a truly successful
project, Brunel went to see the committee, and told
them that he would give them not the cheapest route,
but simply the best.

After much discussion the committee appointed
him as engineer of the Bristol Railway – later to be
called the Great Western Railway. Brunel never
forgot the promise he had made.

In order for a railway to be built in the 1830s a Bill
had to be passed in Parliament giving permission.
Following the success of other railways, like the
Stockton and Darlington built by George and Robert
Stephenson, more and more people came up with
ideas for new lines. Railway mania had arrived. In
the forefront were engineers seeking fame, and financial backers seeking fortunes. Their interests were
not always the same.

Ideas to build a line between London and Bristol
had been voiced in 1825 and 1832, but the first ap-
proach to Parliament did not come until 1834 when
Brunel made his plans. Because money for the line

Great Western Railway,

BETWEEN

BRISTOL AND LONDON.

Capital, £3,000,000, in Shares of £100 each. Deposit £5 per Share.

UNDER THE MANAGEMENT OF A

BOARD OF DIRECTORS,

CONSISTING OF THE

LONDON COMMITTEE.		BRISTOL COMMITTEE.	
JOHN BETTINGTON, Esq.	ROBERT HOPKINS, Jun. Esq.	ROBERT BRIGHT, Esq.	WM. SINGER JACQUES, Esq.
HENRY CAYLEY, Esq.	EDW. WHELER MILLS, Esq.	JOHN CAVE, Esq.	GEORGE JONES, Esq.
RALPH FENWICK, Esq.	BENJAMIN SHAW, Esq.	CHAS. BOWLES FRIPP, Esq.	JAMES LEAN, Esq.
GEORGE HENRY GIBBS, Esq.	HENRY SIMONDS, Esq.	GEORGE GIBBS, Esq.	PETER MAZE, Esq.
ROBERT FRED. GOWER, Esq.	WILLIAM UNWIN SIMS, Esq.	THOS. RICH. GUPPY, Esq.	NICHOLAS ROCH, Esq.
RIVERSDALE W. GRENFELL, Esq.	GEORGE WILDES, Esq.	JOHN HARFORD, Esq.	JOHN VINING, Esq.

C. A. SAUNDERS, Esq. *Secretary.*

Office, No. 17, Cornhill.

W. TOTHILL, Esq. *Secretary.*

Railway Office, Bristol.

Bankers — LONDON :— Messrs. GLYN, HALLIFAX, MILLS, & Co.
BRISTOL :— Messrs. MILES, HARFORD & Co. / Messrs. ELTON, BAILLIE, AMES & Co. / Messrs. STUCKEY & Co.

Solicitors — LONDON :— Messrs. SWAIN, STEVENS & Co.
BRISTOL :— Messrs. OSBORNES & WARD.

Engineer, J. K. BRUNEL, Esq

Applications for Shares to be addressed to the Secretary in London or Bristol, from whom the Prospectus may be obtained.

Subscribers will not be answerable beyond the amount of their respective Shares.

The establishment of a Railway for the important purpose of connecting the Western Districts of this Country with the Metropolis has been resolved upon, as the result of a minute investigation, conducted by the Municipal Corporation of Bristol in concert with the other Public Bodies of that City.

The general advantage of Railways is no longer to be considered a speculative theory. They have invariably conferred an additional value upon the property contiguous to them ; and this fact alone establishes their claim to be considered as National undertakings, contributing to the permanent Wealth of the Country.

The Great Western Railway is recommended also by peculiar local advantages.

The actual amount of travelling on the Line of the projected Railway is even greater than is commonly supposed : it exceeds, in fact, that on any other road to an equal distance from London : it forms the communication of the metropolis and its vicinity Westward, with Windsor, Maidenhead, Reading. Oxford, Cheltenham, Gloucester, Bath and Bristol. Daily communication to a very large extent, takes place by Coaches from the latter places to the West and South-west of England, including the clothing districts of Wilts, Somerset and Gloucester, Wells, Bridgewater, Taunton, Exeter, Plymouth, Devonport, Falmouth, &c.: and by Steam-boats, with the Ports of the Bristol Channel, South Wales and Ireland. On considering these various and extensive sources of Revenue to the Railway, the amount, subsequently stated, will excite no surprise.

The additional facility of intercourse afforded by the establishment of a Railway, will infallibly multiply the present large number of travellers. That mode of communication will also afford an improved conveyance of goods: it will encourage manufactures in the vicinity of the coal fields which surround Bristol, and in both these ways promote the commerce of that Port: it will diffuse the advantages of the vicinity of towns over the whole tract of country intersected by it: it will improve the supply of provisions to the Metropolis, as well as to those towns, and extend the market for agricultural produce; it will also give considerable employment to the laboring class, both during its construction and by its subsequent effects; and will enhance the value of property in its neighbourhood.

was short the first "Great Western Railway Bill" contained the proposal to start at both ends and build lines between London and Reading, and between Bristol and Bath. The company hoped to join the lines in the middle when they had enough money to do so. There was also some opposition to the line. A route was planned to Windsor, but it passed by Eton College, which did not like the idea of noisy dirty trains coming near the school. Because of these difficulties, the Bill was narrowly defeated in the House of Lords.

Soon a second Bill was presented covering the whole line from Bristol to London. This time the Bill was passed and became law on 31st August, 1835. As with the first proposal, it was planned to start the line at both ends and join up in the middle later. There would be branch lines at three points; from Didcot to Oxford, from Swindon to Gloucester, and from Chippenham to Bradford.

Work began from both ends of the line, and the first section of the railway opened was the eighteen miles (29 km) from Acton to Taplow. Taplow was called "Maidenhead" although in fact it was on the opposite side of the Thames from Maidenhead. Apart from the Wharncliffe Viaduct built over the River Brent, the most interesting work on the first section

Below The Wharncliffe Viaduct over the River Brent, the only major engineering project on the first section of the GWR to be completed: the line between Paddington and Taplow (Maidenhead)

was the Skew Bridge over the Uxbridge Road at Southall. This was later damaged by fire. The London terminus was changed from Euston to Paddington. The station was built on the site of the present Goods Depot. This section of the GWR from Paddington to Taplow was opened to the public on 4th June, 1838.

Taplow remained the Western terminus of the line for over a year, while Brunel was building his famous bridge over the Thames at Maidenhead.

This bridge was the centre of furious argument in its early days. The Thames commissioners would not allow any obstruction to the towpath or navigation channel of the river. Brunel in turn did not want to alter the level of his line. These two conditions could only to be met by a low-lying bridge seven and a half metres above the river, with only one river support.

Above The bridge over the Thames at Maidenhead, completed in October 1838

Brunel designed the bridge with two huge, very flat arches 39 metres long and only 7.3 metres high to the topmost point of the arch.

These plans seemed to play into the hands of Brunel's critics, who called the design an extravagant folly. Many of them were sure that the bridge would fall down under its own weight, even before the first train tried to cross it. Never before in the world had anyone tried to build a bridge with such a shallow arch.

In October, 1838, the main supports were taken away and the bridge stood up. One or two bricks dropped out and the critics were sure the rest would follow. But they were wrong. Since that time very much heavier trains have crossed the bridge in perfect safety.

In July, 1839, the line was opened through to Twyford. On the next section of the line Brunel excavated a deep cutting at Sonning, near Reading. After many difficulties, including the resignation of the contractors in the face of an overwhelming sea of mud, the cutting was completed under the direction of Brunel himself. The line to Reading was opened in March, 1840.

In June, 1840, Brunel opened a further section of line. It ran from Reading to Steventon, along the middle reaches of the Thames Valley. Following the river, he took his line through the Goring Gap in the Chiltern Hills. The workings of this section included a cutting at Purley and bridges over the Thames at Basildon and Moulsford.

Below Men at work on the Sonning Cutting, near Reading

Above The elegant Temple Meads
Station at Bristol

Meanwhile work had started on the Bristol to Bath
section of the railway. Because of the uneven lie of the
land in the area, Brunel needed all his engineering
skills to keep the line as level as possible. He had to
dig seven tunnels in the few miles between Bristol and
Bath. In December, 1840, the line was extended
westwards from Steventon to Wootton Bassett and
from here to Chippenham in May 1841.

Now there were only a few miles left between Chippenham and Bath. This was the most difficult section
to build, as almost all of the railway had to be built on
an artificial level. It was raised above valleys on
viaducts and embankments, and cuttings were made
through steep hills. The most difficult work in this
section had already started. This was the digging of a
tunnel through Box Hill. Brunel's plan was for a tunnel
about three kilometres long, with a gradient of one in

Above The famous Box Tunnel portal

a hundred. It was to be the longest railway tunnel yet built, and it attracted a lot of criticism. The engineers of Brunel's time called it "that monstrous and extra-ordinary, most dangerous and impracticable tunnel at Box." Brunel was not put off, and work had begun on the tunnel in September, 1836.

By modern standards, the methods of tunnel construction used by Brunel were primitive and very dangerous. Box Tunnel was built by men and horses, not by machines. All of the spoil was brought to the surface in buckets, pulled up the shafts by horses: 18900 cubic metres in all. Brunel loved to ride in these buckets and tried to get other important people to do the same, but not all of them liked the idea as much as he did! Every week a ton of gunpowder and a ton of candles were used under Box Hill and for two years, up to four thousand men and three hundred horses toiled under the hill. Over one hundred of these men died in accidents on the site.

The tunnel was bored from both ends. In these circumstances, however careful the planning, engineers always keep their fingers crossed that the two tunnels will meet up exactly in line. Brunel was in the tunnel when the break-through came. The two parts of the tunnel were exactly in line. He was so pleased that he gave his ring to the foreman in charge.

Despite delays and difficulties caused by flooding, the tunnel was completed. It was a perfectly straight line. On 21st June, the longest day of the year, it is possible to stand at the west end of the tunnel and see the sun rising through the eastward opening.

In June, 1841, the tunnel was finished, and the line from London to Bristol could be opened. The two cities were linked by a railway of bold design and daring construction. It had been an expensive line too, costing £6,500,000 to build, more than twice the original estimate.

The line was later extended to Exeter, in the West Country, and branch lines were built from Didcot to

Above The GWR at Bath

Oxford, and from Swindon to Gloucester. From Gloucester it was to run along the South Wales coast to Milford Haven.

When Brunel tried to take the line from Gloucester to Cheltenham he ran into opposition. The arguments that followed resulted from a long-standing problem in railway building and one which Brunel had encountered from his first involvement with it — broad gauge or narrow?

Right A front view of the broad-gauge engine *Balaclava*, on Great Western rails in 1892

5 Broad or Narrow?

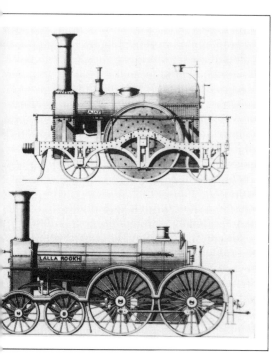

Above Early GWR locomotives from 1837 to 1855

Left The broad-gauge passenger engine *Lightning*, built by the GWR in 1847

When the first railways were built in the North of England, the gauge (that is, the distance between the rails) was 4 feet 8½ inches (1.434 m). This odd distance was used simply because it happened to be the width between the wheels of the early coal wagons in Northumberland. Thus, when the first metal rails were laid in constructing the early railway lines in this area they were set to take the coal wagons of this size. George Stephenson accepted this gauge when he built the Stockton and Darlington Railway, and it finally became the standard for the whole country and much of the world.

In the early days of railway construction, many small private railway companies were formed and they all used this gauge. Locomotives and rolling stock (coaches and wagons) were already being built to fit the narrow-gauge tracks, and it would have been expensive to get special ones made for a different gauge. It was also obvious from early on in the building of the railways, that different companies would have to link up with others close by. Today, we think of railways as a national network, covering the whole country. But they started off as a cluster of separate companies often with less than 80 km of railway to run. These small companies joined up with their neighbours to run trains straight through. This through-running of trains was only possible by joining up railways built to the same gauge. It made sense, therefore, for everybody to pick the same gauge. Almost every railway engineer used this narrow gauge from the very beginning. The exception was Isambard Kingdom Brunel.

Brunel wanted his railway to be not only the fastest in the world, but also the safest and most comfortable. For speed he would need large engines and for safety and comfort he needed a wide track. He chose a gauge of seven feet. On this broad track his trains would run faster and smoother than anyone else's.

The first Great Western Railway Bill, which failed, had referred to the narrow gauge of 4 feet 8½ inches. In 1835, when the second Bill was brought forward, Brunel persuaded Lord Shaftesbury to leave out the clause about the gauge. When the Bill was

Below The last broad-gauge train pulls out of Paddington Station on 20th May, 1892, on its final run to Penzance

safely passed he put forward his idea to the Directors. He based his case on the improved power, speed, space and stability that the seven feet gauge would allow.

It is strange that a man as forward-looking as Brunel did not realize that eventually his own railway would have to link up with others. Perhaps he thought that the others would change to his own broad gauge, once it had proved its worth. If he had been given ten years start this might even have happened. But whatever the reasoning behind his stubbornness to use the broad gauge, his choice was to cause him, and his Locomotive Engineer Daniel Gooch, a great deal of trouble.

Above Daniel Gooch (1816–89) in 1845. At the age of 20, Gooch had been appointed Brunel's Chief Locomotive Assistant on the GWR. From 1865 to 1889 he was Chairman

STEAM FORGING.

BOILER HOUSE.

6 The Locomotive Maker

Brunel's choice of a seven feet gauge caused immediate problems in the building of engines for the Great Western Railway. For some strange reason he laid down very difficult limits for the locomotive designers to follow. He limited the weight of the locomotives to 10.5 tons (10.67 tonnes), and said that any weighing over 8 tons should be carried on six wheels. He also ruled that the pistons of the engine should not travel at more than 280ft/minute (1.42 m/sec) when the engine was travelling at 30 mph (48.28 kmph). This was a very low piston speed. The unfortunate designers had little chance of making good engines if they stuck to these rules. The narrow-gauge locomotives of the day were much heavier than this and their piston speeds were almost twice as fast in many cases. Narrow-gauge locomotives often had 5ft (1.52 m) driving wheels, their pistons had an 18in (45.7 cm) "stroke" and travelled 504ft/minute (2.56 m/sec) at 30 mph. Why did Brunel make these rules? It was probably because he wanted to prevent damage to his track. But he made locomotive designers go to great lengths to meet his rules. Builders were forced to make locomotives with small piston strokes and huge driving wheels. These wheels were so heavy that the designers had to build small boilers on the engines or they would have exceeded Brunel's weight limit. The result was inevitable. A collection of freaks was built; unreliable locomotives without enough power to do their job properly. This was entirely Brunel's fault, but he was

Left Scenes at the GWR's works in Swindon, from the *Illustrated Exhibitor*, 1852

lucky enough to appoint a Locomotive superintendent who saw him over these problems. His name was Daniel Gooch.

Isambard Kingdom Brunel and his Locomotive Engineer Daniel Gooch were very different types. Brunel was unpredictable and brilliant while Gooch was careful and painstaking. Brunel was the inventor, and Gooch the man who shaped Brunel's ideas into practical reality. Together they made a very strong team. Indeed, if it had not been for Gooch, Brunel would probably have lost his job in the first few years of the Great Western Railway.

Daniel Gooch had started work as Locomotive Superintendent of the Great Western Railway on 18th August, 1837, when he was nearly 21. He had already had a lot of experience, including some time with the firm of Robert Stephenson and Company in Newcastle, the makers of the famous *Rocket*. Eighteen locomotives had been ordered for the railway and they were being built to Brunel's absurd specifications. When they arrived, Gooch was in despair. He recorded in his diary:

> "I was not very pleased with the design of the engines ordered. They had very small boilers and cylinders and very large wheels ... I felt very uneasy about the working of these machines, feeling sure that they would have enough to do to drive *themselves* along the road."

The first batch of locomotives was delivered from the Vulcan Foundry. These were *Vulcan*, *Aeolus* and *Bacchus*. Similar locomotives, *Apollo*, *Neptune* and *Venus* were delivered later. By the end of 1838, the Directors of the Great Western Railway had become so indignant because their locomotives kept breaking down, that they demanded a special report from Gooch.

He said that *Venus* was one of the worst engines

Right Sketch of the GWR locomotive *Vulcan*

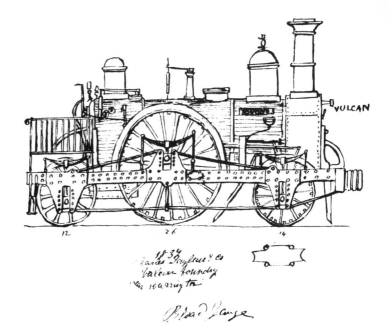

Below Photograph of *Vulcan*. This
engine was one of the first three to
be delivered from the Vulcan
Foundry – in 1838

No 150
R Stephenson & Co
Engineers
Newcastle upon Tyne
1837

Left Sketch of the GWR locomotive *North Star,* one of two which Daniel Gooch had designed for Robert Stephenson & Company of Newcastle, for export to Russia. They were never exported and in 1837 Gooch persuaded Brunel to acquire them and convert them to broad gauge

ever delivered. He had been unable quite to make any use of her as a regular train engine. He had been forced to keep her on shunting duties, out of harm's way. In this report Gooch had stressed the faults of workmanship, but stopped short of blaming Brunel for the weaknesses in the design. Later he pointed this out, and drew a rebuke from Brunel, but Brunel saw the truth of what Gooch was saying. Later he supported Gooch's recommendations for alterations to the engines, although in the event only *Venus* and *Apollo* were altered.

Other locomotives were made by Mather, Dixon and Company of Liverpool and by Sharp, Roberts and Company. These too, were all unreliable engines.

When the locomotives had first been delivered Gooch had realized that they were not powerful enough to draw coaches and wagons with any reliability. He remembered that while he was working with Robert Stephenson he had designed two engines himself. Gooch bought them for the Great Western Railway and converted them to the broad gauge.

The first of these engines was the famous *North Star*. The other, which came later, was the *Morning Star*. These locomotives were of the normal design and not built to meet Brunel's design specifications.

North Star cost £2,150 and weighed 18.75 tons (19.05 tonnes) much heavier than Brunel's recommendation of 10.5 tons. It proved to be the only reliable locomotive of the first bunch, and must have done a lot to prove to Brunel that he had made a mistake. The broad gauge flyers that later established the supremacy of the Great Western Railway and justified the broad gauge were direct descendents of this locomotive. Before Gooch could earn fame with these locomotives, however, he received a lot of unfair criticism.

Even though he had shown that he had an unreliable collection of badly designed and badly built locomotives, some of the blame rubbed off on him. There

Left Photograph of *North Star*. Although bought by the GWR in November 1937, *North Star* had to await, covered in tarpaulin, the arrival of her sister engine *Morning Star* before starting in service

is an entry in G H Gibbs' diary of 1838:

"Our engines are in very bad order, and Gooch seems to be very unfit for the superintendance of the Locomotive Department."

Realizing that Daniel Gooch was not responsible for the poor performance of the early locomotives, Brunel let him make his own designs. Gooch prepared these with his usual care, and produced the first example of standard design. While making new designs, Gooch ordered some more locomotives of the *Star* class from Stephenson. *Evening Star* and *Dog Star* were early arrivals. Daniel Gooch ordered ten altogether, to add to the two he already had, *North Star* and *Morning Star*. The "Stars" were not exactly alike, but similar in design to each other.

Gooch's own passenger engines were directly based on Stephenson's *North Star*, but they were better locomotives. Once again Gooch displayed his talent for improvement and development. Without Gooch's locomotives, Brunel's idea for a broad gauge railway of superior quality would have remained a splendid dream. It was to Gooch and engines of the quality of *Ixion* and the *Iron Duke* class that Brunel owed a great deal.

Below The *Iron Duke*, built at the GWR Works, Swindon, and completed in April 1847. With engines of this class and power, the time of the fast service from Paddington to Exeter fell to 4 hours 25 minutes

Meanwhile, Brunel's broad gauge was running into other problems. While the whole of the line from London to Exeter and South Wales – and its branch lines – had been built at a gauge of seven feet, other railway companies had used the narrow gauge. When the branch to Gloucester was built, it met up with a line coming the other way from Birmingham. After a heated argument the two companies agreed to share a mixed-gauge line between Gloucester and Cheltenham. This was a poor settlement of the problem. Something had to be done to decide which gauge was to become the standard, and in 1846 the government appointed a Royal Commission to do this.

Brunel felt that the battle of the gauges should be decided on merit. He suggested a sporting contest, a sort of Olympic Games for locomotives. This idea was accepted and it was agreed that trials should be held between Paddington and Didcot on the broad gauge, and between Darlington and York on the narrow gauge.

Daniel Gooch picked *Ixion* as his champion locomotive. Stephenson picked a new locomotive called "Engine A" for his. The contrast between the names of these locomotives is interesting. The romantic *Ixion* against the boring "Engine A". It seemed to symbolize the battle of the gauges. *Ixion* produced 60 mph (96.6 kmph) with an 80-ton load, while Engine A could not exceed 54 mph (86.9 kmph) with 50 tons. A second locomotive provided by the narrow-gauge supporters became derailed after 35 kilometres. The commissioners reported that the trials "confirm the statements given by Mr Gooch".

Brunel and Gooch had won the contest. But how important was the contest? Brunel was still acting as though the argument was about technical achievement alone. It was not. Two factors were heavily against them. Firstly the broad gauge could only boast of 440 km of railway compared with the 3059 km of narrow gauge. Standardization could

Overleaf The Engine House, Swindon

only reasonably go one way. Secondly the narrow gauge was cheaper to build. Although they recognized the superior speed of the broad-gauge engines the Royal Commissioners had no alternative but to recommend the narrow gauge as the standard British Railway gauge. Brunel and Gooch did not give up the struggle, however.

Daniel Gooch designed and built *Great Western* at Swindon in 1846. This locomotive had a large 8ft (2.44 m) single driving wheel, but boiler and cylinders were in proportion. In 1847, *Great Western* was altered to give it two leading axles. It eventually completed 370,000 miles. Gooch improved his design further and created a super locomotive. This was the first of the famous *Iron Duke* or *Great Britain* class of 8-foot single driving wheel locomotives. These became the standard express locomotives of the Great Western Railway until the broad gauge was abolished in 1892. They were bigger, better, and more powerful than any other locomotive of the day.

In 1848 Gooch displayed his practical flair once again by producing the first *dynamometer car* to measure traction performance (this is really a measurement of the efficiency of the locomotive – how effectively it pulls the rolling-stock given the size of its driving wheels and boiler etc). Even today, traction performance is measured by techniques based on Gooch's early example.

Although Gooch designed bogie locomotives (these have sets of swivelling wheels which are able to turn with the track) and tank engines, (ie. locomotives that carry water and coal on board and not in a tender) it will be for the *Iron Duke* class that he will always be remembered. He started a tradition of Great Western steam supremacy which led, generations later, to the famous *Castle* and *King* class locomotives which many people think were the best express steam locomotives ever built. Wherever steam locomotives are discussed, the name of Daniel

Gooch will be remembered. He was among the first of the truly great designers.

Yet the battle for the gauge had already been lost. Although a few more miles of broad-gauge railway were laid, the narrow-gauge builders had firmly established themselves all over the country. By refusing to co-operate with plans to build stretches of mixed gauge rails, they had stopped all opposition to the narrow gauge. Yet Brunel had made his point: he and Daniel Gooch had created a new generation of locomotives. Their speed became a legend, and their effect on the growth of the railways continued until steam was replaced by the diesel engine.

Below Poem, by a sympathizer, on the death of the broad-gauge railway

THE BURIAL OF THE "BROAD-GAUGE."

MAY 23, 1892.

["Drivers of Broad-Gauge Engines wandering disconsolately about with their engine-lamps in their hands; followed by their firemen with pick and shovel over their shoulder, waiting in anxious expectation of the time when that new-fangled machine, a narrow-gauge engine, should come down a day or two after."—*Times' Special at Plymouth on Death of Broad Gauge.*]

Not a whistle was heard, not a brass bell-note,
  As his corse o'er the sleepers we hurried;
Not a fog-signal wailed from a husky throat
  O'er the grave where our "Broad-Gauge" we buried.

We buried him darkly, at dead of night,
  The sod with our pickaxes turning,
By the danger-signal's ruddy light,
  And our oil-lamps dimly burning.

No useless tears, though we loved him well!
  Long years to his fire-box had bound us.
We fancied we glimpsed the great shade of BRUNEL
  In sad sympathy hovering round us.

Few and gruff were the words we said,
  But we thought, with a natural sorrow,
Of the Narrow-Gauge foe of the Loco. just dead,
  We should have to attend on the morrow.

We thought, as we hollowed his big broad bed,
  And piled the brown earth o'er his funnel,
How his foe o'er the Great-Western metals would tread,
  Shrieking triumph through cutting and tunnel.

Lightly they'll talk of him now he is gone,
  For the cheap "Narrow Gauge" has outstayed him,
Yet BULL *might* have found, had he let it go on,
  That BRUNEL's Big Idea would have paid him!

But the battle is ended, our task is done;
  After forty years' fight he's retiring.*
This hour sees thy triumph, O STEPHENSON;
  Old "Broad Gauge" no more will need firing.

The "Dutchman" must now be "divided in two"!—
  Well, well, they shan't mangle or mess *you!*
Accept the last words of friends faithful, if few :—
  "Good-bye, poor old Broad-Gauge, God bless you!"†

Slowly and sadly we laid him down.
  He has filled a great chapter in story.
We sang not a dirge—we raised not a stone,
  But we left the "Broad Gauge" to his glory!

* The Royal Commission appointed to inquire into the uniformity of railway gauges, presented their report to Parliament on May 30, 1846.
† Words found written on one of the G.-W. rails.

TO A DEAR YOUNG FEMININE FRIEND, WHO SPELT "WAGON" AS "WAGGON."

BAD spelling? Oh dear no! So tender, she Wished that the cart should have an extra "*gee.*"

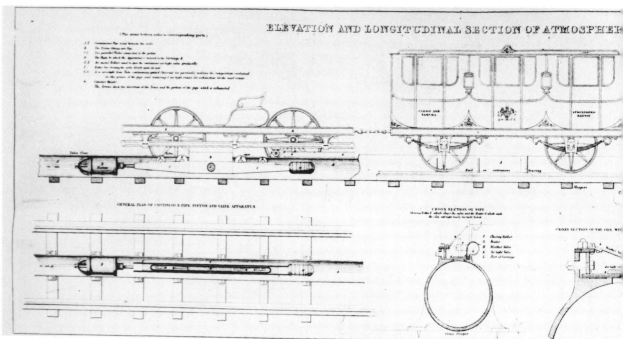

ELEVATION AND LONGITUDINAL SECTION OF ATMOSPHERIC

GENERAL PLAN OF CONTINUOUS PIPE PISTON AND VALVE APPARATUS

CROSS SECTION OF PIPE

CROSS SECTION OF THE PIPE WITH

7 The Atmospheric Railway

Left South Devon Atmospheric Railway pumping station at Totnes

Below Design for an atmospheric railway, as envisaged by the Samuda brothers and Samuel Clegg

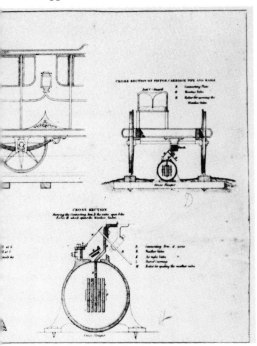

For most of his life Brunel had more than his share of success. This success was often more enjoyable because people kept on saying he would fail. Although some of his achievements were controversial, his career was largely successful. One of Brunel's projects, however, was an outstanding failure. This was the "atmospheric railway", the novel system of travel he adopted in South Devon. The idea behind the "atmospheric" railway was simple. Instead of having trains pulled by locomotives, they were pushed forward by air pressure. To do this, a hollow tube or pipe holding a piston was laid down between the rails along the line. The leading coaches were joined on to the piston, and when air was sucked out of the tube, the piston would be forced forward and the train pulled along. The pumping was done by pumping stations nearly every five kilometres along the line.

This idea was not as silly as it sounds. One advantage of the atmospheric railway was that trains could be built much lighter, since they did not require a good grip on the rails (adhesion) to provide drive (i.e. the forward push from the wheels). The system was a mechanical way of doing exactly what is done on electrified railways today. A diesel engine or a steam engine makes its own power as it goes along. But an electric locomotive does not make electricity. It only uses it. The electricity comes to the locomotive from the power station through overhead wires or along the third rail. The pumping stations of the atmospheric railway can be compared with the electric

61

power stations of today. The tube between the rails was like the overhead wires. It brought the power to the train. This system was really anticipating the electrification of railways; the separation of the power source from the locomotive. But perhaps it was too far ahead of its time. The mechanical systems available at the time could not overcome the difficulties the system involved. Electrical systems would have done so, but on the other hand, engineers cannot afford to be too far in front of their time. An engineer has to *make* things. He is limited by what can be done with the methods and materials of his day.

When Brunel was made engineer to the South Devon Railway, he decided on a fast broad-gauge line linking up with the Great Western Railway through Exeter and Bristol. The broad-gauge locomotives were doing very well over the fast and level line between Paddington and Bristol. So why did Brunel want to change the system between Exeter and Plymouth? He had recently seen a trial demonstration of an atmospheric railway, and he thought he could use the idea to solve the special problems of South Devon. These problems were caused by steep hills. The line from London to Bristol was very fast because it was straight and level for most of the way. It would have cost a fortune to build a level railway in the hills of South Devon. So Brunel wanted to find some way of going fast uphill. The "atmospheric" system looked as if it might be the answer. Since all of the drive is supplied by the wheels of the locomotive, steam engines on steep gradients with heavy trains can easily develop "wheel slip". This is when the wheels spin around without getting a grip on the rails and is the reason why railways are always built as near level as possible. Roads can usually follow the ups and downs of the countryside without any great difficulty.

The first suction system was invented in 1810 by George Medhurst, but the first experiment did not

Above Old corroded pipes from the Atmospheric Railway, lying in a Paddington yard in 1958

take place until 1840, in West London. It was known as "Mr. Clegg's pneumatic railway". Various inventors made improvements to this system. It was demonstrated that a load of 11 tons (11.8 tonnes) could be moved at 22 mph (35.4 kmph), powered by the suction in a 9in (23 cm) pipe. The power came from a fixed steam-engine in a pumping house. All the well-known railway builders took a look at the system. Stephenson thought it was "humbug". Brunel thought it was a great idea.

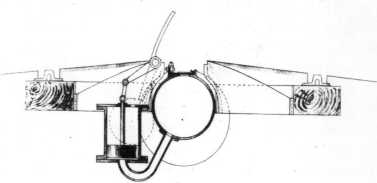

Atmospheric Railway.
Sketch of Bridge, Counterpoise, and Vacuum Cylinder
for a

Road crossing the Mains, on their level.

Position of Bridge &c when the Main is exhausted and a Train approaching

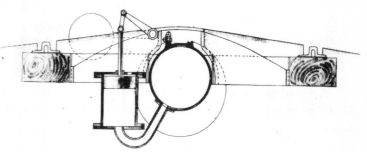

Position of the Bridge &c. when the Main is not exhausted.

Right Cross-section design of the Atmospheric Railway pipes

Scale 1 inch to 1 foot

In 1843, another test track was opened, this time in Ireland, on the Kingstone to Dalkey branch near Dublin. It proved a risky business. The line was up-hill, 2.4 kilometres long. The suction pipe was 38 centimetres across, and a huge pumping engine was built at Dalkey at the top of the hill. Trains were held back by brakes but the driver did not have as much control as he would have liked. A fatal accident nearly occurred when the front coach was not coupled onto the rest of the train and was propelled at a very high speed. This incident showed once again the major snag in the suction railway. The train was not under

Above Map section and view of the
South Devon Atmospheric
Railway at the St Thomas station

proper control all of the time. You could put on the brakes, but how could you switch off the engine when it was in a pumping station three or four kilometres away?

Brunel knew about these problems, but he thought he could beat them. He reported to the South Devon Railway saying that the "mechanical difficulties can be overcome". His first pipes were put down between Exeter and Teignmouth, and later as far as Newton Abbott. Brunel built pumping stations every three miles (4.83 km) – at Exeter, Countess Weir, Starcross, Dawlish, Teignmouth, Summer House, and Newton Abbot.

But when the line to Teignmouth and Newton Abbot was opened in 1846, it was Gooch's engines that were called in to do the job. The pipe system was in trouble. It had cost a lot more than had been thought and had also run into technical problems. To allow the link between the piston inside the pipe and the train it pulls, the pipe had to have a slot all the way along the top. Since the pipe was supposed to be airtight (for pumping air out in front of the train) the pipe had to be fitted with a leather flap which could only open as the train passed. This never worked properly from the beginning. The pumping engines were also unreliable, partly due to Brunel's idea that they should work at high pressure.

Finally, these problems were overcome, and in September, 1847, the great day came. Trains were working between Exeter and Teignmouth. "Mr. Clegg's pneumatic railway" had arrived. In January, 1848, the "atmospheric" system was operated as far as Newton Abbot. But this triumph did not last. The system could not stand up to the wear and tear of use. There were many minor accidents. Water got into the pipes. The leather "valves" split and the pipes were leaky. The pumping stations could not keep the pressure going without heavy fuel costs. The pipes got worse. In winter the leather froze stiff. By June,

1849, before the railway had been working even one year on the new system, the leather along the whole line was useless. It was rotting away. Brunel had a choice. He could pay £25,000 for new leather (and have the same trouble again after a few months) or he could give up. He was brave enough to admit he had made a mistake. He gave up.

In 1848, the pumping stations and equipment were sold. It was the end of a dream, but not the end of the line. Gooch's locomotives were called back into service, and not for the next hundred years was the supremacy of the steam locomotive seriously challenged on the main line to the West.

Despite the failure of the "atmospheric railway" in South Devon the GWR eventually went as far as Devon and Cornwall, to the most south-westerly extreme of the country. This further extension of the line led to the building of the Royal Albert Bridge over the River Tamar at Saltash. This was one of Brunel's successes.

The first major obstacle facing anyone building a railway from Plymouth into Cornwall is the River Tamar. At Saltash, where Brunel planned to cross this river, it was over 300 m wide. In the middle, the water was 20 m deep. It was also tidal. The Navy wanted at least 30 m clearance beneath the bridge, so they could sail ships up the Tamar beneath it.

Brunel decided to use the method he had used at Chepstow to carry the Gloucester to South Wales line over the River Wye. He would use arched tubes for the main spans, and suspend the railway beneath these on chains, rather like a suspension bridge. At one time Brunel toyed with the idea of 4 spans, and he even considered trying a huge, single span. In the end he decided on two spans, with a deep water pier in the middle of the river to support them. Brunel decided his two main spans would each be 142 m. To save money he planned a single-line railway over the bridge. Trains from both directions would use the

Above The Royal Albert Bridge at
Saltash, under construction, with
one main truss complete and in
position

Above The Saltash bridge under construction

same track.

In building the bridge, the biggest problem was the deep-water pier in the middle. A second problem was how to get the huge tubes into position for the main spans.

Brunel used a cylinder for the central pier. It was made on the spot and floated out on the tide. This huge iron cylinder was 37ft (11.3 m) in diameter and 85ft (25.9 m) high. It could be used as a diving bell. At the bottom of the river was $3\frac{1}{2}$ m of mud, and beneath that rock. Brunel was able to get a foundation on this firm rock and build the pier up to water level.

The big wrought iron tubes were also built on the spot. They were oval in cross-section 16ft (4.9 m) high and 12ft (3.66 m) across. They were arched and spanned 465ft (142 m). These large spans were floated into position and swung at high tide over the base of the piers. They were then lifted by hydraulic jacks and the piers built up below. The bridge got

Above A hundred years on, the Cornish Riviera Express pulls away from Brunel's single-line bridge onto double tracks

higher and higher. Every week it grew about two metres as the piers were built.

The bridge was completed in July 1858 – 730yd (667.5 m) long, with two main spans and seventeen side spans on curved approaches. The bridge cost a quarter of a million pounds to build.

The story of the final stages of the bridge construction is a sad one. In September, 1857, the tricky "floating day" brought huge crowds to Saltash. In complete silence Brunel gave his orders from a position high up on the structure. It was Brunel the showman once again. When success was achieved, there was much applause. A band struck up "See the conquering hero comes". The failures of the atmospheric systems were forgotten. But even at this moment of triumph, time was running out for Brunel. Although he might not have realized it, he was a dying man. This was the last time he would receive the acclaim of the crowd, the last time he would complete a major work.

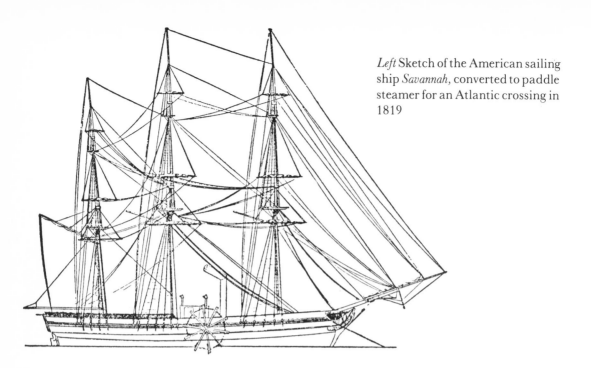

Left Sketch of the American sailing ship *Savannah*, converted to paddle steamer for an Atlantic crossing in 1819

8 *The Race to New York*

Brunel was a genius of restless ambition. He was determined not to be just "the railway man". With the details of the London to Bristol railway fixed in his mind another dream was taking shape. This was trans-Atlantic steam navigation. At a meeting of the Great Western Directors doubts were expressed over the London to Bristol railway line. It seemed such a huge enterprise. Brunel got to his feet. "Why not make it larger and have a steamboat go from Bristol the New York and call it the *Great Western?*" he demanded. His suggestion was greeted with nervous laughter. Only Isambard's friend Thomas Guppy, a Bristol merchant, took the idea seriously.

Brunel's move into steamship design was in spite of a warning by Marc. Marc himself had been asked to become engineer for a steamship company serving the West Indies. He wrote, "As my opinion is that steam cannot do for distant navigation, I cannot take part in any scheme." The problem then confronting the trans-Atlantic ships has recently been echoed by the supersonic aircraft *Concorde*. How could they carry sufficient fuel for long journeys? Although in 1819, the American ship *Savannah* had crossed the Atlantic, it was really a sailing ship with paddles which could be taken out of the water.

A Great Western Steamship Company was formed at Bristol. Typically, Brunel's first attempt was to be a world record vessel, larger than any previous ship, and built of oak. The sternpost was set up on 28th

Left The *Great Western* leaves Bristol for New York. She was the first steamship to run a regular service across the Atlantic and completed seventy-four crossings in the course of her career

July, 1836, by William Patterson, who was to build the ship to Brunel's design. Very soon afterwards Bristol was the scene of a clash between Brunel and the eccentric scientist who had challenged Brunel and Gooch in the great locomotive contest, Dr Dionysius Lardner. Lardner was speaking to the British Association on the subject of trans-Atlantic steam navigation. At the end of the lecture Brunel leaped to his feet and challenged the "expert's" facts and figures but Lardner remained firm. Most of the audience could not follow the debate between them which quickly became very technical. Brunel's response was typical of the man. Actual performance was the best proof. Dionysius Lardner might spout in theory about the impossible. Isambard Brunel would simply *show* them!

In July, 1837, the *Great Western* floated out of Patterson's dock into Bristol harbour and left for London under sail to have her engines fitted. The *Great Western* was a majestic ship. Its huge hulk was painted a sinister black, brightened by the gold Neptune and twin dolphins at the bows. But there were those who were less than pleased at the splendid sight! London and Liverpool saw themselves in direct competition with Bristol. In those days steamship companies behaved like first division football clubs today, and the competition between them was intense. London and Liverpool were building rival boats that would not be ready in time. London's *British Queen* certainly had no hope of being the first steamship to New York. But London were not beaten yet! They had chartered a little vessel called *Sirius* from the St George Steam Packet Company and had already increased her coal bunker space. It was the *Sirius* that slipped away first, bound for New York, on 28th March, 1838. It was not until the last day in March that Brunel's *Great Western* moved off in pursuit. The battle of David and Goliath had begun.

The thundering motors and thrashing paddle

Above The St George's Steam Packet Company's *British Queen* on her maiden voyage from London to New York in July 1839. She made only nine Atlantic crossings and was broken up in 1842

Right The *Sirius*, which beat the *Great Western* to New York by one day

wheels of the *Great Western* left a long, white wake astern as she tried to get back into the race. Then it happened. Flames and smoke belched out from a boiler room and suddenly Captain Claxton and Brunel were fighting the dread of all mariners, a fire at sea. Brunel, thinking of all the coal on board, was worrying about firefighting when he trod carelessly on the burned rung of a ladder and fell into the boiler room. Only by landing on the unfortunate Claxton did Brunel escape certain death. Even so, he had to be wrapped in a sail and carried painfully on to Canvey Island where the *Great Western* had been run aground. The damaged ship was repaired at Bristol and did not get away until 8th April, 1838.

Meanwhile Roberts, captain of the little *Sirius* had taken on coal at Cork and was battling out a freak Atlantic storm. In spite of pleadings from passengers and crew alike, he pressed on grimly for New York. After three thousand miles the *Sirius* began to run out

of coal, just as Dr Lardner had forecast. But rumour had it that the fanatical Roberts simply burned cabin panelling and furniture in place of coal. By this means he arrived at New York, to a tumultuous welcome on 22nd April. Brunel's vast, oak steamship came in close behind the *Sirius* and its size and majesty turned many heads. For many years the *Great Western* held the coveted Blue Riband for the fastest crossing, but nobody could detract from the glorious first crossing made by the little *Sirius*. Within two years of this triumph, however, the heroic Roberts was to go down at the helm of another ship in a mid-Atlantic storm.

The *Great Western* went on to record sixty-seven successful crossings in eight years. By any account the *Great Western* was the better ship, a genuine trans-Atlantic steamboat. The *Sirius* win had been a freak result, but Brunel himself was forced to recognize the courage of her captain.

Above The *Great Western,* run aground on Canvey Island after she caught fire in her first attempt to leave Bristol for New York at the end of March 1838

Right IK Brunel – from a photograph taken in the last year of his life

THE GREAT BRITAIN, IRON STEAM SHIP, OF BRISTOL.

This immense Iron Steam Ship, which has excited the most lively interest in all connected with naval architecture, was built at the yard of the Great Western Steam Ship Company, Bristol, under the superintendence of Thomas R. Guppy, Esq. C.E. No expense has been spared in her construction, and whether we consider her tonnage, or the grand problem she is calculated to solve,—that of the best method of propelling, and the material to be preferred in the construction of large steam vessels,—the Great Britain is confessedly the most splendid experiment in ship building ever submitted to a British public. Constructed wholly of iron in her outward framing, her capacity for stowing goods is much greater than in wooden vessels; whilst in rate of sailing, it is confidently anticipated, she will distance all competitors. She has six masts, as shewn in the Engraving, and she can therefore avail herself to the fullest extent of a favourable wind; but as they are considerably lower than the masts in ships of much smaller size, they will offer very little resistance when steaming against adverse winds.

The weight of iron used in the Ship and Engine is upwards of 1500 tons. By the action of low pressure steam, used expansively, in four cylinders of 88 inches in diameter, with 6 feet stroke, 1000 horse power will be applied to a Propeller of 15 feet in diameter, revolving under the stern,—an improvement on the Archimedean Screw, which it is expected will supersede the cumbrous paddle wheel. The following are the dimensions of this magnificent Ship:—

Length from Apron head to taffrail	322 Feet
Length on Upper and Forecastle Decks	308 Feet
Main breadth	50 Feet 6 Inches
Depth	32 Feet 6 Inches
Promenade Cabin, forward	Breadth, 31 Feet 9 Inches	Length	67 Feet
Ditto ditto aft	Ditto, 24 Feet	Ditto	110 Feet
Dining Saloon, forward	Breadth, 21 Feet 9 inches	Ditto	61 Feet
Ditto ditto aft	Ditto, 30 Feet	Ditto	98 Feet 6 Inches

22 State Rooms with one Bed; 113 ditto with two Beds.

Published by J. C. Kingdon, St. James's Square Avenue, Milk Street, Bristol.

9 *The Big Ships*

Brunel did not rest on his laurels, but almost immediately began to plan an iron steamship. He got the idea after seeing a little iron paddle-steamer called the *Rainbow* at Bristol Docks. Brunel was never afraid of borrowing ideas, but when he did he enlarged them in scale or scope. Brunel decided to call it the *Great Britain*. It was to be the biggest steamship in the world. Like his father before him, Brunel put his full trust in only a few select people and so he hoped that Maudslay Sons and Field would build the paddle engines. But the lowest tender for building the engines was given by Francis Humphreys who saw it as a chance to make a name for himself. But, imagine Humphreys' bitter disappointment when Brunel suddenly abandoned paddles and switched to screw propulsion. This was because Brunel had seen the little screw-driven ship *Archimedes* in the port of Bristol during May, 1840. The blow finished off the sensitive Humphreys who wasted away in despair. But Brunel's decision left him with a highly original steamship. Never before, it seemed, had so many novelties been combined in a single vessel. On 19th July, 1843, the great iron, screw-driven *Great Britain* was launched by the Prince consort.

In January, 1845, the *Great Britain* was in the Thames being fitted out. She was graced with a visit from the Queen. As it was unthinkable for Her Majesty to go below deck, Brunel showed her the action of the engines with a working model! But the

steamship was cursed. It wasn't long before news
arrived of her going aground in Southern Ireland.
Her captain, Hosken, had no idea where she was, and
thought he must have hit the Isle of Wight! This
raised a question that not even Brunel had thought
of. Do iron ships disturb magnetic compasses? It
was a worrying thought and everybody was relieved
to discover that Hosken's charts had been faulty.

But Brunel's huge schemes were no longer serving
the interests of Bristol. The Docks authorities had
dithered on the alterations needed to the harbour and
locks. It soon became clear that the big ships would
have to dock elsewhere. Like the seven foot gauge, the
Great Western Steamship Company was destined to be
a magnificent failure. The company had lasted only
ten years, earned nobody a fortune, and had only
built two ships. It seemed that Isambard Kingdom
Brunel was interested only in fame and not in fortune.

The subsequent history of the *Great Britain* is full of surprise and romance. In February, 1886, she left Penarth for Panama. On rounding the Horn, it was discovered that her cargo of coal was on fire. The captain struggled back to the Falkland Islands where she was beached like a giant whale. For a time the huge hulk was used for storing wood and coal but was eventually towed away to rest in peace at Sparrow Cove. In 1969, she was salvaged and towed to Port Stanley in the Falkland Islands. After this she was returned to the Port of Bristol, where she is now being restored. It is hoped to fit her up as a maritime museum.

Below Run aground in Dundrum Bay on the north eastern coast of Ireland, the *Great Britain* is being protected from further damage by "a mass of strong faggots lashed together with iron rods . . . like a huge poultice"

To Brunel, the biggest steamships in the world were challenges to make even bigger ships. In the same way that every fisherman dreams of hooking and landing a "monster", so Brunel's thoughts turned to "the big one". Brunel's sketch book shows that thoughts of "the big one" had been with him for some time. Alongside drawings of the new Paddington Station, there suddenly appears a quite stunning sketch of a steamship, bristling with funnels and masts. Brunel labelled the sketch *East India Steamship* and adds, almost as an afterthought, its dimensions, "say 600ft. × 65ft. × 30ft." These were figures to make a contemporary engineer catch his breath. Six times the volume of anything afloat!

Brunel's idea was eventually taken up by the Eastern Steam Navigation Company. This company was in effect supporting the project on money that Brunel

Below In 1969, the *Great Britain* was towed for restorative work to Port Stanley in the Falkland Islands where she had run aground in 1886, seeking shelter from a heavy storm

himself had rallied. Those in the company who liked the idea of "hooking a monster" wanted to call the steamship *Leviathan*. Brunel preferred to keep the family of names that bound his ships together and suggested *The Great Eastern*. In fact, the project had moved East, to Napier Yard on the Thames, near the Isle of Dogs. The launching itself was to be a controlled sideways launching. This meant that the ship was built parallel to the river.

The building of the *Great Eastern* was marred for Brunel by his collaboration with Scott Russell. This engineer had been made responsible for making part of the steamship, but unfortunately he had tried to avoid his responsibilities and Brunel had been forced to denounce him to the company. But for the help of Brunel's friends, led by Gooch, the whole project would have been disastrous.

Right John Scott Russel (1802–82), the Scottish engineer who was heavily involved in the building of the *Great Eastern,* but whose behaviour and incompetence almost killed the project and ruined its backers

10 *The End of the Line*

Left Chain-drums and checking gear used at the first attempted launch of the *Great Eastern*

Below An exterior and a cut-away view of the *Great Eastern*

The launching of the *Great Eastern* was possibly the biggest technical challenge of Brunel's career. A sideways launching isn't a question of brute force. It needs great care. Experiments suggested to Brunel that the launching cradles would best slide down an inclined plane of one in twelve. But for once, Brunel was too cautious.

The *Great Eastern* took a very long time to launch. Brunel was severely criticized after the ludicrous "launching ceremony", but it wasn't his fault really. Just about everything had gone wrong and a second launching was necessary. The press accused Brunel

of crude showmanship and he was made to feel like a playwright given bad first-night reviews. But Brunel did not want a highly publicized launching. For a successful launching the actions of many people needed to be co-ordinated. In the absence of modern public address systems, quiet was essential. Instead, Brunel arrived at Napier Yard to find a carnival atmosphere. The company secretary had sold over three thousand tickets. The launching was as crowded as Nottingham Goose Fair. Later, Brunel wrote to the Directors, "I learned to my horror that all the world was invited to the launch." It also rained most dismally. And finally, there was the tragedy of an unnecessary death. An elderly worker was killed by the handles of a spinning winch. This unnerved the launching team and Brunel wisely abandoned the attempt to launch the ship. Even the name was wrong. As the bottle broke she was named *Leviathan*, although she was eventually registered as the *Great Eastern*. It was by that name she became widely known. *The Times* continued that paper's hostility to

Below Richard Tangye (1833–1906) beside one of the Tangye hydraulic rams used for the sixth – and successful – launching of the *Great Eastern*

Above Tensely, four men watch a renewed attempt to launch the great ship. They are, left to right, John Scott Russel, Henry Wakefield (one of Brunel's assistants), IK Brunel and Lord Derby

Brunel with a quiet sneer: "We seem to have been a little unfortunate in our grandiose schemes of late." *The Times* thought Brunel was only interested in his own fame and saw him as a bad financial risk.

Eventually, over many weeks, Brunel increased the hydraulic power until the great ship was actually jacked all the way down to the river. Among the collection of borrowed and built equipment producing the power were some presses made by the Tangye brothers. This up-and-coming firm was subsequently able to boast, "We launched the *Great Eastern* and she launched us."

Sadly, the effort was too much for Brunel. He was already ill. Working at all hours on damp winter nights had sapped his remaining reserves of energy. By the time the *Great Eastern* was afloat it was obvious to all that Isambard Kingdom Brunel was dangerously ill. After a winter in the sun had failed to restore him to health, Brunel returned to England and directed his last desperate energies towards fitting out the ship for its maiden voyage. Often he would dictate reams of instructions from his sick-bed and his weary, gaunt figure would occasionally be seen stumbling about his own ship. Sometimes his thoughts would flicker on to possible new ventures. But it was all too late. The spirit was being forced more and more to submit to the weakness of the flesh.

Grittily, Brunel hung on for one final triumph, the sailing of the *Great Eastern*. But it was never to come. The *Great Eastern* sailed with one fatal flaw. A turned-off steam-cock allowed steam pressure to build up beneath the decks. This was not a design error and

Left The *Great Eastern* at New York

Left Brunel aboard the *Great Eastern*

Right The wrecked and torn funnel jacket after the boiler room explosion

easily avoidable. But that day, and in those conditions, disaster was impossible to avert. The great ship was passing Dungeness Light when the huge, forward funnel exploded upwards into the blue sky. Instantly, the paddle boiler room filled with scalding steam. People like the atomic-bomb victims of Hiroshima staggered out. Friendly hands grabbed them by the arm only to find the skin come away from the flesh as if they were steamed fish. One ran screaming overboard to be crushed immediately by the giant paddle wheel. The happy event had, in a second, become a major disaster.

Bravely they bore the grim news to the dying Brunel and in his last hours the famous engineer was forced to taste once again the bitterness of defeat. But the *Great Eastern* went on to prove her worth, and completed a dazzling career by laying the first trans-Atlantic telegraph cable. Gooch went on to become a baronet and was close to the *Great Eastern* in many of her triumphs. But perhaps his greatest satisfaction, which he called an "omen of success", was when coming into Cork harbour the *Great Eastern* was met by a little tug. It was called simply by a single name, *Brunel*.

The Characters

BENTHAM, Sir Samuel (1757–1831)   A naval architect and engineer who became Inspector General of Naval works.

BREGUET, Abraham Louis (1747–1823)   The famous French maker of watches, clocks and precision instruments. The young Brunel was for a time apprenticed to this fine craftsman.

BRUNEL, Isambard Kingdom (1806–1859)   The subject of this book. The designer of the Clifton Suspension Bridge, Brunel is perhaps best known for his work as engineer to the Great Western Railway, and as the builder of transatlantic steamships.

BRUNEL, Sir Marc (1769–1849)   Born in France, the father of Isambard Brunel worked as an engineer in New York before settling in England. He was a pioneer in machining block pulleys. His major works include Chatham dockyard. His Thames tunnel was completed in 1843.

CLEGG, Samuel (1781–1861)   An inventor and gas engineer who pioneered the suction system which became known as "Mr Clegg's pneumatic railway".

GIBBS, George (1785–1842)   One of the first Directors of the Great Western Railway. He kept a diary which provides us with some detailed information about the early years of the GWR.

GOOCH, Sir Daniel (1816–1889)   The famous English engineer who became the first Locomotive Superintendant of the Great Western Railway. He designed many famous engines for Brunel's broad gauge. He later laid the first Atlantic cable. In 1866 he was made a Baronet.

GUPPY, Thomas (1797–1882)   A Bristol merchant. One of the early sponsors of the scheme to build a railway between London and Bristol.

KINGDOM, Sophia (1776–1854)   Marc's wife and Isambard's mother. It was from her that he got his middle name "Kingdom".

LARDNER, Dr Dionysius (1793–1859)   Sometimes portrayed as a comic figure – but really a considerable authority, popularizer and writer on science. He clashed with Brunel over the effect of wind resistance on locomotives and the fuel consumption of ships.

MAUDSLAY, Henry (1771–1831)   A well-known engineer who with Joshua Field (1757–1863) founded the famous engineering firm of Maudslay, Sons, and Field.

Left The famous photographic portrait of Brunel by the chains of the stern checking drum used to control the *Great Eastern*'s launch in 1857

STEPHENSON, George (1781–1848) An English locomotive inventor who became a famous railway engineer. In 1829, his locomotive *Rocket* won a series of trials. He was engineer to the Stockton and Darlington Railway which was built in the 4ft 8½ins gauge, later to become standard for the whole country. In 1826, he was appointed engineer to the Liverpool and Manchester Railway.

STEPHENSON, Robert (1803–1859) The son of George Stephenson, Robert worked with his father on a number of railway projects. He also designed bridges.

TELFORD, Thomas (1757–1834) The famous engineer and early bridge builder. Telford built bridges over the River Severn at Montford and Buildwas. He planned the Ellesmere and Caledonian canals. His extensive works include over 1600 kilometres of road and over 1200 bridges. Among the best known are the London–Holyhead road, the Menai Suspension Bridge (1825) and St Katherine's Docks, London. He criticized Brunel's design for the Clifton Suspension Bridge, and submitted a design of his own. He was the first President of the Institution of Civil Engineers.

TREVITHICK, Richard (1771–1833) The Cornish mining engineer and inventor Trevethick designed steam carriages and high pressure steam engines. He was called in to help when early efforts at a Thames Tunnel ran into problems.

VAZIE, Robert (1771–1837) Mining and tunnel engineer, known as "the Mole". Vazie proposed a Thames Tunnel as early as 1802.

Date Chart

1753	Bridge-building fund set up in Bristol for a bridge over the Clifton Gorge.
1769	Marc Brunel born in Normandy, France.
1793	Marc Brunel flees from the French Revolution and goes to America.
1799	Marc Brunel settles with Sophia in England.
1802	Robert "the Mole" Vazie plans the first Thames Tunnel. The Thames Archway Company is formed.
1806	Birth of Isambard Kingdom Brunel.
1808	Thames Archway Company folds up as project runs into difficulties. Trevithick's tunnel collapses.

1814	Marc Brunel's Battersea Sawmills destroyed by fire.
1816	Birth of Daniel Gooch.
1820	Isambard Brunel goes to school in France.
1821	Marc Brunel arrested and imprisoned for debt.
1823	Isambard Brunel joins his father's office.
1824	Thames Tunnel Company formed.
1825	Work starts on the Brunels' Thames Tunnel.
1827	The Brunels hold a banquet under the Thames.
1828	Work stops on Thames Tunnel.
1829	A Bridge Committee is formed in Bristol.
1830	Competition for best design of Clifton Bridge. Isambard Brunel wins.
1831	Work starts, briefly, on Clifton Suspension Bridge.
1832	Meeting at Bristol to discuss the possibility of a London–Bristol railway.
1833	Isambard Brunel made engineer of the Great Western Railway.
1833–4	Brunel surveys the route for the new railway.
1834	First Great Western Railway Bill put to Parliament and rejected.
1835	Second Great Western Railway Bill put to Parliament and passed.
	Brunel's broad-gauge proposals accepted by GWR.
	Work starts again on the Thames Tunnel.
1836	Work starts again on Clifton Suspension Bridge.
	Work starts on the *Great Western* steamship at Bristol.
	Brunel is married.
	Work begins on Box Tunnel.
1837	Daniel Gooch is made Locomotive Superintendent of the GWR. First locomotives delivered. The *Great Western* steamship is launched at Bristol.
1838	First section of the GWR opened from Paddington to Taplow.
	The *Sirius* makes the first transatlantic steamship crossing.
	The early locomotives prove unreliable.
	Maidenhead Bridge is completed.
1839	Maidenhead Bridge is opened and the line extended to Twyford.
1840	Line extended to Farringdon Road. Work starts at the Bristol end of the Bristol–Bath section.
	Gooch chooses Swindon as the "principal engine establishment" of the GWR.
	"Atmospheric railway" is demonstrated in West London and seen by Brunel.

1841	Box Tunnel completed. Final section of the line between Bath and Chippenham is opened.
	Marc Brunel is knighted.
	The Thames Tunnel is opened.
1843	Demonstration of the suction railway system on the Dalkey branch near Dublin.
	The *Great Britain* steamship is launched at Bristol.
1844	GWR line opened as far as Exeter, and the Oxford branch opened.
1845	The *Great Britain* is fitted out in the Thames.
1846	Gooch's locomotive the *Great Western* arrives, the first of many locomotives from the Swindon works. The line extends further west, to Teignmouth and Newton Abbot in South Devon.
1847	The Exeter–Teignmouth stretch of the line worked on the "atmospheric" system.
1848	Gooch invents a "dynamometer car" to measure traction performance. The atmospheric system is extended to Newton Abbot.
1849	Atmospheric system fails and has to be abandoned. Death of Sir Marc Brunel.
1852	Work begins on one of Brunel's masterpieces, the new Paddington Station.
1853	The *Great Eastern* steamship is started. Work begins on the Royal Albert Bridge over the Tamar at Saltash.
1854	Paddington Station complete.
1858	Launching of the *Great Eastern*.
1859	Death of Isambard Kingdom Brunel. In the same year Robert Stevenson and Dionysius Lardner also die.
1864	Opening of Clifton Suspension Bridge.
1886	Daniel Gooch becomes a baronet.
1889	Death of Sir Daniel Gooch.
1892	The broad gauge is abolished.

Glossary

CIVIL ENGINEERING   A branch of engineering concerned with designing and constructing roads, railways, buildings, bridges, tunnels and earthworks.

FRENCH REVOLUTION   An uprising of the French working people against their ruling classes in the late eighteenth century.

ROYALIST   Supporters of the Monarchy at the time of the French Revolution.

REPUBLICAN   Supporters of the uprising against the Monarchy at the time of the French Revolution.

BATTLE OF WATERLOO   A famous battle in 1815 in which the French were defeated by the rest of Europe.

DRIFTWAY   A preliminary tunnel of narrow diameter sometimes used in the construction of larger tunnels.

THAMES BARGE   A flat-bottomed vessel for plying freight traffic on the Thames Estuary between the larger ships and the various wharves. Early barges had sails. Later barges were pulled along by tugs.

SQUIRES' COACHES   Horses and carriages belonging to the gentry.

CITY MERCHANTS' DRAYS   Horses and carts for delivering goods traffic around London.

EXCREMENT   Raw sewage tipped into the Thames.

GAS CANDELABRA   Large ornate light fittings with many sources of light and lit by gas.

FLORID GOTHIC   A very ornate and decorative architectural style of a kind usually found in churches and important public buildings.

RAILWAY MANIA   A wave of enthusiasm for railway building in the mid nineteenth century when everybody believed that there was money to be made in building new lines.

THE GORING GAP   A gap in the Chilterns Hills where the river Thames crosses the range. Railway builders looked for natural gaps to avoid expensive tunnels.

PISTON SPEED   The speed at which the piston passes into the cylinder of a steam engine, measured in number of strokes per minute.

DRIVING WHEELS   The wheels of a locomotive which are coupled to the piston and so provide the forward thrust on the rails.

SUPERINTENDANCE   Being the manager of a Department.

DYNAMOMETER CAR   A railway vehicle which contains special equipment by which the performance of the locomotive in pulling the train can be measured.

PNEUMATIC OR ATMOSPHERIC RAILWAY   A railway which is designed to move trains forward by means of suction in a continuous tube containing a piston to which the train is attached. Suction is provided by pumping stations along the line.

PADDLE STEAMER   A boat which is driven forward by large "paddle wheels" turning in the sea and not by a screw propeller.

MARITIME MUSEUM   A museum containing items of interest from naval or seafaring sources.

FIRST NIGHT REVIEW   A report given in a newspaper after a new play has opened in a theatre, which many use as a guide to see if the play is worth seeing or not.

Further Reading

The best and most valuable source for further reading is LTC Rolt's *Isambard Kingdom Brunel* (Longmans, Green, 1957) most readily available in paperback form in the Pelican Biographies Series.

Other sources include:
Brunel, Isambard, *The Life of Isambard Kingdom Brunel, Civil Engineer,* (Longmans, Green, 1870)
Ellis, C Hamilton, *British Railway History* (Allen & Unwin, 1954)
Ellis, C Hamilton, *The Pictorial Encyclopedia of Railways* (Hamlyn, 1968)
Marshall, John, *The Guinness Book of Rail Facts and Feats* (Guinness Superlatives Ltd. 1975)
History of the Great Western Railways Vol 1 & 2 (GWR, 1927–31)
Nobel, Lady Celia Brunel, *The Brunels, Father & Son* (Cobden Sanderson 1938)
Spratt, HP, *Outline History of Transatlantic Steam Navigation* (London HMSO, 1950)

Index

Acknowledgements

The authors and publisher thank the following for their kind help with illustrations and permission for them to appear on the following pages:—
   Associated Press 80; British Railways 15, 47, 60–61, 62–63, 64–65, 68, 69; Brunel University Library 34, 84, 75, 86, 88; Mrs MV Brunel Hurst, 8, 11; National Maritime Museum 12, 81; National Railway Museum 35 & 36, 39, 42; Science Museum (Crown Copyright) 10, 16, 18, 19, 21, 26, 32, 34, 37, 38, 40–41, 43, 44–45, 46–47, 51, 52, 54, 56–57, 66–67, 70, 72–73, 75, 76, 79, 83, 87; Victoria & Albert Museum 14; Wellington Museum 13.
The remaining pictures belong to the Wayland Picture Library.